MW00936630

Design Control and Manufacture
of
Medical Devices
for Engineers

Priscilla Browne

ISBN: 9781090217783

Contents

Forward

The aim of the short book is to provide an understanding of the importance of design controls in device quality and safety for the patient and end user. Design controls interact with man elements of a companies quality management system and they have a continual role in post market surveillance and maintaining the product design throughout its lifecycle.

Design Control and their statutory regulations ensure that good quality management (QM) practices are used for the design of medical devices and products remain consistent with quality management systems

Design controls increase the probability that the design transferred to production will result in a medical device that performs and functions as intended and meets the user's needs.

Providing a safe and effective medical device is critical for the success of any firm or manufacturing company. This book covers the nine main areas of design control listed below. It is an ideal desktop companion or resource for those new to design controls or those impacted by them.

Design and development (D&D) planning
Design input(s)
Design output(s)
Design review
Design verification (Design V&V)
Design validation
Design transfer
Design changes(managed under change control)
Design History File (DHF)

Introduction

Design controls are a collection of practices and procedures that are incorporated into the design and development (D&D) process for a regulated product such as a medical device.

Design controls are required per FDAs Quality System Regulation, 21 CFR Part 820. They apply to a large variety of medical devices with varying levels of complexity and application (implants, delivery systems, fixation, transient use)

The regulation itself does not prescribe the practices that must be used, rather it works to establish a framework that manufacturers must use when developing and implementing design controls. Such requirements should be appropriate to ensure that regulation allows design controls to be flexible enough to meet individual manufacturers own design and development processes. Manufacturing and design of medical devices is a diverse area which covers simple, single use devices, over-the-counter devices and more complex devices used by professional medical practitioners.

Based upon quality assurance principles and quality by design, the application of GMP and good engineering practice (GEP)- design controls provide a structure and clear path from user needs (intended use) to product delivery through a step-by-step standardised process. As the word "design" suggests-design controls ensure safety and quality right from the beginning.

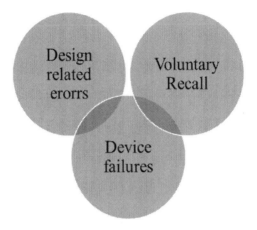

Fixing a design issues as soon as they arise in development reduces the cost of doing so at a later point and ensures the resultant design is appropriate for its intended use.

Bringing a formal review process (design control) to the table assists engineers and managers in engaging with decisions and understanding them better. It also ensures that when future changes are made, they are documented and reviewed adequately with proper consideration to the design inputs.

Design controls are a requirement of quality systems such as 21 CFR Part 820 (medical devices), and for certain classes of devices and per ISO 13485 - Quality Management Systems.

FDA regulations requires manufacturers of all Class III and Class II devices to apply design control.

Benefits of Design Control:

- The intended use of the device (product) is documented and approved by appropriate
- It ensures input(s) align with output(s)

- Problems with designs or manufacturability are recognised earlier
- It creates a design "standard" and a "process" to allow benchmarking/ consistency within an organisation

The importance of the design input(s) and verification of design outputs is fundamental to good Design control processes. When the design input has been reviewed and the design input requirements are determined to be acceptable and formally documented, an iterative process of translating them into a device and design begins to unfold. The first step is conversion of the requirements into system or high-level specifications. Thus, these specifications are a design output. Upon verification that the high-level specifications conform to the design input requirements, they become the design input for the next step in the design process, and so on.

This basic technique is used repeatedly throughout the design process. Design inputs are converted into a design output. Following this, each output(s) is verified as conforming to its input(s).

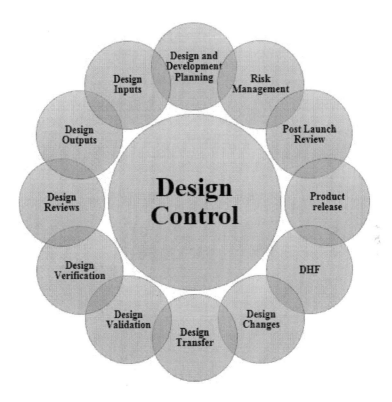

Concurrent Engineering

Mostly, a concurrent engineering approach is adopted during the design processes used in the industry.

Concurrent engineering allows for a team to work together and efficiently to maintain control of a design but also allow greater flexibility to respond to changes and development and design improvement. In practice, this can often lead to some cross over between design been "locked down" and the start of production. Therefore, concurrent models need correct quality and management oversight.

Design Controls and ISO 13485 Quality Management System

Clause 7 of ISO 13485 (Medical Device Direction) specifies the requirements for design and development (D&D) of devices as part of the product realisation process. It should be noted that organisations can opt to exclude specific requirements of ISO 13485, in cases where product realisation is not applicable. However, any such exclusion should be based on sound rationale with the technical case clearly documented and approved by stakeholders. An example of this may be where design and development activities are not conducted by the manufacturer e.g. contract manufacturers.

Clause 7 (product realisation) of ISO 13485 details requirements for design and development controls. Clause 7 includes the following subparts:

Clause 7.1 Planning of product realisation
Clause 7.2 Customer-related processes
Clause 7.3 Design and development
Clause 7.4 Purchasing
Clause 7.5 Production and service provision
Clause 7.6 Control of measuring devices

Section 7.3 (Design and development) comprises:

Clause 7.3.1 Design and Development Planning
Clause 7.3.2 Design and Development Inputs

Clause 7.3.3 Design and Development Outputs
Clause 7.3.4 Design and Development Review
Clause 7.3.5 Design and Development Verification
Clause 7.3.6 Design and Development Validation
Clause 7.3.7 Control of Design and Development Changes

Definitions

Change Management: a management process where changes to the product, process, facilities or utilities are assessed, planned and reviewed as part of a formal systematic process.

Corrective and Preventative Action (CAPA): when an unplanned or adverse event happens, a corrective and preventative action can be implemented.

Design and Development Plans
The DHF must include a design and development plan and any project schedule.

Design Phase Review: a process of evaluating the design requirements against the ability of it to deliver the intended device.

Design History File (DHF): an approved list of records that describe the design history of a medical device. At the beginning of Design control a DHF should be created and then maintained. It should detail the design and development activities that relate to all aspects of the device such as the device itself, materials, components, labelling and packaging and production methods.

Design History Record (DHR): a record that contains the production history of a manufactured finished device.

Design Input: the physical and performance requirements of a device that are the basis for the device design. Design inputs need to eventually trace to the design outputs and where tested- verification and validation (V&V).

Design Output: the results of a design effort at each design phase and at the end of the total design effort. The finished design output is the basis for the device master record. The total finished design output consists of the device, its packaging and labelling, and the device master record.

Design Verification: confirmation by examination and provision of objective evidence that specified requirements have been fulfilled.

Design Transfer Documentation: The DHF should include design transfer documentation showing that the device design was correctly translated into production specifications.

Design Validation: establishing by objective evidence that device or product specifications conform to user needs and intended use(s) defined in design documentation.

Device Master Record (DMR): a compilation of records containing the procedures and specification for a device. The contents of a DMR can contain local procedures such as SOPs and work instructions along with global or divisional specifications used to detail manufacturing processes, intermediate product or final product.

Design Phase Review(s): a documented, comprehensive, systematic examination of a design to evaluate the adequacy of the design requirements, the capability of the design to meet those requirements and to identify problems.

Intended use: the use of a product, service or process in as per the instructions for use or information provided by the manufacturer.

Independent Reviewer: A technically knowledgeable person without direct responsibility for the design under review.

Post Launch review: after the launch of a new product, a post launch review is completed to monitor its safety and effectiveness and its risk and complaint performance.

Specification: specification means any requirement to which a product, process, service, or other activity must conform.

User Needs: knowing the intended use of a medical device allows the designer to then identify user needs. The device must be safe and effective for the user (Patient or Medical professional)

Validation: validation means confirmation by examination and provision of objective, documented evidence that the particular requirements for a specific intended use can be consistently fulfilled.

Application of Design Controls

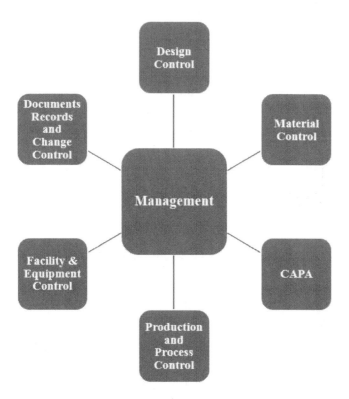

Design controls can be applied to any product development process. However, the application of design controls for certain classes of medical devices is mandatory. When the design input has been reviewed and the design input requirements are determined to be acceptable, the process of creating the device design begins. Design control procedures should distinguish between the research phase and the development phase. As the design develops. product specifications are drafted and compared to the design input requirements.

In turn, the product specification becomes the input for the next step in the design process. In the development and drafting of product specifications (e.g. critical quality attributes etc.) review of applicable product standards and industry best practices such as ISO and ASTM bodies should be reviewed. For example, a catheter manufacturer should develop products with reference to ISO 10555 - intravascular catheters - sterile and single and the product must conform to the requirements therein.

The Phase Approach to Design Control

The term "phase approach" is often used when describing the design control process. It simply means that a sequence of tasks needs to be completed, reviewed and approved during the development cycle of a product or medical device. Tasks are grouped into phases or stages. At the beginning, technical issues relating to design input requirements may need to be addressed with solutions identified. Often a range of solutions can be available, utilising different technologies. These different solutions then go on to be reviewed at the design selection process. At design selection, the project team must choose and justify a particular solution. The next phase (such as design verification and validation) ensures that the design is transferred to product launch and commercial supply - no oversights or deviations in the design intent occur. It also ensures that the device meets the user needs and intended uses (design inputs).

A phase review is a process of evaluating the progress against the goals and activities of a particular phase. The phase review is typically completed at the end of each phase, but there may be a need to complete interim reviews for long or complex projects. For example, a design phase review is completed to ensure that the design input requirements make sense before they are interpreted into design specifications (design input phase review).

Risk Management

Risk management involves the systematic application of management policies, practices and procedures that identify, analyse, control and monitor risk.

It is important to recognise that risk management should begin at the outset of the design and development phase of a project. The first step is to identify the user needs and intended use and application of the device. At the design input phase and design selection phase, risk assessments should be in a mature state. This allows the review of potential risks relating to the design of the product. Unacceptable risks can be dealt with by means of revisiting the design or introducing controls or mitigations in order to reduce the risks to acceptable levels. Following on from the design and development phase, the design verification, validation and transfer phases, or the clinical readiness phase, risk management activities and acceptability of the residual risk become the focus and must be approved indicating acceptability. This is often referred to as communicated risk.

In order to apply a risk management strategy, a procedure or SOP on risk management is typically available within manufacturing companies. This should clearly describe the risk management process and the various risk assessment tools, their application and guidance on how to complete them. The content of any risk management procedure or SOP should align with ISO 14971:2007 Medical Devices - Application of Risk Management to Medical Devices. Controlled templates for PFMEAs etc. also bring consistency and continuity to the process.

Risk management begins with the development of the design input requirements. As the design evolves, new risks may become evident. To systematically identify and, when necessary, reduce these risks, the risk management process is integrated into the design process. In this way, unacceptable risks can be identified and managed earlier in the design process when changes are easier to make and less costly.

An example of this is an exposure control system for a general-purpose x-ray system. The control function was allocated to software. Late in the development process, risk analysis of the system uncovered several failure modes that could result in overexposure to the patient. Because the problem was not identified until the design was near completion, an expensive, independent, back-up timer had to be added to monitor exposure times.

Risk Control

Where risks are identified as unacceptable, risk control measures must be determined to reduce the risk prior to the process or system being implemented. A number of actions can be taken in order to further reduce risk including: (1) changing the design to reduce risk, introducing protective measures in the device or the manufacturing process, (3) inserting a warning statement into the instructions for use (IFU).

Risks scored as "investigate further risk reduction" should be examined to determine whether it is practicable to reduce the risk further. The risk should be reduced to as low as is reasonably practicable, (aka ALARP) taking into account the benefits of accepting the risk and the practicability of implementation. If risks classed as "investigate further risk reduction" are already at ALARP, no further risk reduction is necessary.

Residual Risk Evaluation

After risk control measures are applied, a new risk assessment will be carried out to determine residual risks. Residual risks will be assessed for acceptability using the same criteria as detailed in 6.5. If the residual risk is not judged acceptable then further risk control measures will be applied.

If the residual risk is not judged acceptable and further risk control is not practicable then the team may perform a risk/benefit analysis by evaluating data and literature on the medical benefits of the intended use to determine if they outweigh the risk. If this evidence does not support the conclusion that the medical benefits outweigh the residual risk, then the risk remains unacceptable. This analysis should be recorded and approved by both the risk management team and senior site management.

EN ISO 14971: 2009– Characteristics - Annex C

Annex C contains several questions that can be used to identify medical device characteristics that could impact upon safety:

1) What is the intended use and how is the medical device to be used?

2) Is the medical device intended to be implanted?
3) Is the medical device intended to be in contact with the patient or other persons?
4) What materials or components are utilised in the medical device or are used with, or are in contact with, the medical device?

EN ISO 14971:2009 - Annex E

Refer to EN ISO 14971:2009 Annex E - Examples of hazards, foreseeable sequences of events and hazardous situations.

Examples of hazards, foreseeable sequences of events and hazardous situations:

Some Examples of energy hazard(s)

Electromagnetic energy
Line voltage
Leakage current
- enclosure leakage current
- earth leakage current
- patient leakage current
Electric fields and/or magnetic fileds
Radiation energy
Ionising radiation
Non-ionising radiation
Thermal energy
High temperature
Low temperature
Mechanical energy

Gravity

Examples of biological and chemical hazards

Bacteria
Viruses
Other agents (e.g. prions)
Re- or cross-infection
Chemical
Exposure of airway, tissues, environment or property, e.g. to
foreign materials:
- acids or alkalis
- residues
- contaminates
- additives or processing aids
- cleaning, disinfecting or testing agents
- degradation products
- medical gases
- anaesthetic products
Biocompatibility
Toxicity of chemical constituents

Some Example of operational hazard(s)

Function
Incorrect output or functionality
Incorrect measurement
Erroneous data transfer
Loss or deterioration of function

Use error
Attentional failure
Memory failure
Rule-based failure
Knowledge-based failure
Routine violation

Examples of information hazards

Labelling
Incomplete instructions for use (IFU)
Inadequate description of performance characteristics
Inadequate specification of intended use
Inadequate disclosure of limitations

Operating instructions
Inadequate specification of accessories to be used with the
medical device
Inadequate specification of pre-use checks
Over-complicated operating instructions

Warnings
Of side effects
Of hazards likely with re-use of single-use medical devices

Examples of initiating events and circumstances

Incomplete requirements
Inadequate specification of:
- design parameters
- operating parameters
- performance requirements
- in-service requirements (e.g. maintenance, reprocessing)

end of life

Manufacturing processes
Insufficient control of changes to manufacturing processes
Insufficient control of materials/materials compatibility
information
Insufficient control of manufacturing processes

Insufficient control of subcontractors

Transport and storage
Inadequate packaging
Contamination or deterioration
Inappropriate environmental conditions

Environmental factors
Physical (e.g. heat, time)
Chemical (e.g. corrosions, degradation, contamination)
Electromagnetic fields (e.g. susceptibility to electromagnetic disturbance)
Inadequate supply of power
Inadequate supply of coolant

Cleaning, disinfection and sterilisation
Lack of, or inadequate specification for, validated
Procedures for cleaning/disinfection/sterilisation
Improper cleaning or disinfection.

Disposal and scrapping
No information or inadequate information provided
Use error

Formulation
Biodegradation
Biocompatibility
No information or inadequate specification provided
Inadequate warning of hazards associated with incorrect formulations
Use error

Human factors

Potential for use errors triggered by design flaws, such as

- confusing or missing instructions for use
- complex or confusing control system
- unclear device state
- unclear presentation of settings, measurements or other information
- misinterpretation of results
- insufficient visibility, audibility or tactility
- poor mapping of controls to actions, or of displayed information to actual state
- controversial modes or mapping as compared to existing equipment
- untrained personnel
- insufficient warning of side effects
- inadequate warning of hazards associated with re-use of single-use medical devices
- incorrect measurement and other metrological aspects
- incompatibility with consumables/accessories/other medical devices

slips, lapses and mistakes

The Quality System and Design Controls

In addition to procedures and work instructions necessary for the implementation of design controls, policies and procedures may also be needed for other determinants of device quality that should be considered during the design process. The need for policies and procedures for these factors is dependent upon the types of devices manufactured by a company and the risks associated with their use. Management with executive responsibility has the responsibility for determining what is needed.

Polices are typically more top level in nature which present the core approach to a subject matter. Example of topics for which policies and procedures may be appropriate are:

risk management	configuration management
device reliability	compliance with regulatory requirements
device durability	device evaluation
device maintainability	clinical evaluations
device serviceability	document controls
human factors engineering	use of consultants
software engineering	use of subcontractors
use of standards	use of company historical data

Design and Development Planning

It is the manufacturer's responsibility to establish and maintain plans that describe or reference the design and development activities and define responsibilities for implementation. The plans should identify and describe the interaction with different groups or activities that are part of the design and development process. The maintenance of plans to reflect an accurate state as the design and development progresses is also a key factor. The design and development planning is intended to be prospective in nature. It allows risks to be identified earlier and promotes timely delivery of projects. Above all, it ensures that all key design and development tasks are addressed during the design and development process

Design and Development Planning Objectives

• Describe the goals and objectives of the design and development project (i.e. what is to be developed).
• Definition and documentation of responsibilities.
• Identification of the major tasks and deliverables. Assign individual or organisational responsibilities

Human factors are user characteristics such as behaviour, physical limitations, cognitive capabilities, and the way a user would logically use a device. Human factors should consider the user population, the patient, and the operating environment.

The FDA considers potential human error to be a design flaw; therefore, considering it at the early stages of development can reduce the potential for device malfunction, injury, or even death.

Process Inputs and Outputs

Inputs
- Product Concept
- Patient requirements
- Regulatory Requirements
- Human Factors Engineering

DESIGN CONTROL

Outputs
- Specifications
- IFU
- Design History File
- Devie Master Records
- Validations

Design Input Phase

The aims of the Design Input Phase are to (1) define and document the user needs and the intended use of the medical device and (2) translate user needs and the intended use of the packaged device into design input requirements. (E.g. engineering specifications and the product requirements specifications.) The typical documents required when establishing design inputs include:

The creation of a formal design description detailing the intended use, user requirements and design inputs. (Note: the design description must align with the design input requirements.)

- A design and development plan which provides an estimation of timelines, resources required, roles and responsibilities, project risks and scope of the project.
- Initial risk assessment which contains the user, design and component risks to be mitigated.
- Design concepts and technology overview e.g. charter or proposal.
- Business case report addressing the market size and market opportunity- also known as a marketing requirements specification.

Examples of design input(s) include

(a) **User requirement:** the device must be sterile:
    a. Sterilisation via terminal sterilisation (e.g. ETO)
        i. The device material may not be compatible with ETO, therefore another method of sterilisation such as Autoclaving via clean steam.

(b) **User requirement:** Device is portable:
    a. Device must not be so heavy it cannot be

easily carried.
    i. Quantify weight with a suitable

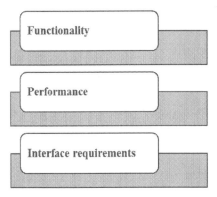

FDA 21 CFR Requirements – Design Input

21 CFR Part 820.30(C) Design Input per FDA.

- *"Each manufacturer (of medical devices) shall establish and maintain procedures to ensure that the design requirement(s) relating to a device are appropriate and address the intended use of the device, including the needs of the user (medical professional) and patient.*
- *The procedures shall include a mechanism for addressing incomplete, unclear ambiguous, or conflicting requirements.*
- *The design input requirements shall be documented and shall be reviewed and approved by designated individuals."*

Incomplete requirements can have a serious effect on the design and ultimate success of a product. If essential design requirements are omitted in error or otherwise, the impact on quality or functionality may not be detected until verification of validation. When it comes to design controls. The intention of term "validation" differs from that used for process validation or equipment validation (IQ/OQ/PQ). Validation when it comes to design control often indicates some sort of in vivo testing using clinical trials. So when an omission of a design requirement occurs, this presents an expensive problem that may not be easily rectified. If design requirements are missed, a redesign may be necessary before a design can be released to production, thus causing delays to the project. An impact assessment of any redesign must determine if any repeating of verification or validation is required. Furthermore, if modifications are required to tooling, or process equipment, timelines can be impacted greatly. However, the safety and quality of the product must be paramount. Keeping one eye on the user requirements and intended use of the product is an important factor in avoiding gross design requirement failings.

What Is Design Input?

An artist's impression or concept documents do not meet the true intent of design input requirements. The purpose of design input is to create a *complete* set of requirements that are written in a technical manner with an engineering and scientific level of detail. The use of qualitative terms in a concept document is both appropriate and practical. This is often not the case for a document to be used as a basis for design. The language used in the creation of Design input(s) and the design stage also has a profound impact on the direction and scope of a product. If a concept document describes the product to be suitable for "outside use", then there will be requirements with regards to insulation, water ingress testing and operating temperatures and so on.

Design input requirements must be comprehensive. This may be quite difficult for manufacturers who are implementing a system of design controls for the first time with little or no experience in the industry of medical devices. Design input(s) requirements fall into three categories with most products having requirements within all three categories including:

(1) Functional requirements detailing the operation of the device-how it should operate
(2) Performance requirements detailing the performance requirements or expectations of the device in relation to accuracy, speed of response times, battery life, device safety and reliability etc. – Performance must ensure the intended use is fulfilled.
(3) Interface requirements specifying features of the device which are critical to compatibility with external systems such as the patient interface- interfaces can present as a keyboard, a download port or a display screen.
The scope of design input(s) work depends on the complexity of a device and the risk associated with its use.

Tips for Reviewing Design Input Requirements

The ultimate goal of the design input phase is to gain agreement and approve the requirements formally. At this point, the document is a controlled document (approved with signatures) and subject to change control. Any updates required at a later date will need to be done through the change control process.

Design Input Requirements Should Be Crystal Clear: For example, a medical device may require use of a built-in battery. It would be important to specify the life expectancy of the battery. To say it has an approximate operating life of 2-3 years is too vague. A better description would be to say it has 2000 hours of operation with a software requirement that logs the number of hours the device is powered on. This mitigates the likelihood of failure during use.

Use of Tolerances: For example, a contact lens may have an outer diameter of 14.00mm. While this is the target/nominal value it cannot be ever accurately achieved. There will always be a degree of variation in the diameter measurement. Applying a tolerance, allows an acceptable range in which the measurement is within specification and accepted. If the diameter is specified as 14.00±0.2mm, designers have a basis for determining how accurate the manufacturing processes have to be. In addition, the specification will allow designers determine if the design meets the intended use.

Industry Standards: Design input requirements should meet or exceed industry standards. Compliance to product specific standards should be considered.

Environment: The operating environment of the device should be specified. Take the example of a cardiac defibrillator. If the device is intended for use on a frontline ambulance it may be used outdoors in cold and damp conditions. On the other hand, use within a hospital setting would require greater control of the temperature range and environmental conditions.

Design input is the starting point for product design. The requirements which form the design input establish a basis for performing subsequent design tasks and validating the design. Therefore, development of a solid foundation of requirements is the single most important design control activity. Many medical device manufacturers have experience with the adverse effects that incomplete requirements can have on the design process. A frequent complaint of developers is that "there's never time to do it right, but there's always time to do it over." If essential requirements are not identified until validation, expensive redesign and rework may be necessary before a design can be released to production.

Human Factors: Human factors are the study of the interactions between humans and device (i.e., interface) and the subsequent design of the device-human interface. It plays an important role in Design Control.

Design Inputs Checklist

- Device functions
- Physical or material characteristics
- Performance in terms of speed accuracy etc.
- Safety to user and patient
- Reliability
- Standards
- Regulatory requirements
- Human factors
- Labeling & packaging
- Maintenance
- Sterilization (is re-sterilisation required or is it single use)
- Compatibility with other devices/products
- Environmental limits

BE CLEAR

UNAMBIGOUS

SPECIFIC

APOPT INDUSTRY STANDARDS

Research and Development

Some manufacturers have difficulty in determining where research ends and development begins. Research activities may be undertaken in an effort to determine new business opportunities or basic characteristics for a new product. It may be reasonable to develop a rapid prototype to explore the feasibility of an idea or design approach, for example, prior to developing design input requirements. But manufacturers should avoid falling into the trap of equating the prototype design with a finished product design. Prototypes at this stage lack safety features and ancillary functions necessary for a finished product and are developed under conditions which preclude adequate consideration of product variability due to manufacturing.

What is the scope of the design input requirements development process and how much detail must be provided? The scope is dependent upon the complexity of a device and the risk associated with its use. For most medical devices, numerous requirements encompassing functions, performance, safety, and regulatory concerns are implied by the application. These implied requirements should be explicitly stated, in engineering terms, in the design input requirements.

There are many cases when it is impractical to establish every functional and performance characteristic at the design input stage. But in most cases, the form of the requirement can be determined, and the requirement can be stated with a to-be-determined (TBD) numerical value or a range of possible values. This makes it possible for reviewers to assess whether the requirements completely characterize the intended use of the device, judge the impact of omissions, and track incomplete requirements to ensure resolution.

For complex designs, it is not uncommon for the design input stage to consume as much as thirty percent of the total project time. Unfortunately, some managers and developers have been trained to measure design progress in terms of hardware built, or lines of software code written. They fail to realize that building a solid foundation saves time during the implementation. Part of the solution is to structure the requirements documents and reviews such that tangible measures of progress are provided.

Assessing Design Input Requirements For Adequacy

Eventually, the design input must be reviewed for adequacy. After review and approval, the design input becomes a controlled document. All future changes will be subject to the change control procedures. Any assessment of design input requirements boils down to a matter of judgment.

Design input requirements should be unambiguous. That is, each requirement should be able to be verified by an objective method of analysis, inspection, or testing. For example, it is insufficient to state that a catheter must be able to withstand repeated flexing. A better requirement would state that the catheter should be formed into a 50 mm diameter coil and straightened out for a total of fifty times with no evidence of cracking or deformity. A qualified reviewer could then make a judgment whether this specified test method is representative of the conditions of use.

Quantitative limits should be expressed with a measurement tolerance. For example, a diameter of 3.5 mm is an incomplete specification. If the diameter is specified as

3.500±0.005 mm, designers have a basis for determining how accurate the manufacturing processes have to be to produce compliant parts, and reviewers have a basis for determining whether the parts will be suitable for the intended use.

The set of design input requirements for a product should be self-consistent. It is not unusual for requirements to conflict with one another or with a referenced industry standard due to a simple oversight. Such conflicts should be resolved early in the development process. The environment in which the product is intended to be used should be properly characterized. For example, manufacturers frequently make the mistake of specifying "laboratory" conditions for devices which are intended for use in the home. Yet, even within a single country, relative humidity in a home may range from 20 percent to 100 percent (condensing) due to climactic and seasonal variations. Household temperatures in many climates routinely exceed 40 °C during the hot season. Altitudes may exceed 3,000 m, and the resultant low atmospheric pressure may

Design Output Phase

The purpose of the design selection(output) phase is to provide a range of design options and solutions with the relevant evidence to show the effectiveness of the same. Often proof of concept (POC) or proof of principle (POP) trials may be used to verify effectiveness of solutions. POC/POP testing can involve making some limited prototypes. Any documents created in the previous phase, design input, should be reviewed and updated if required. There should be no contradictions or gaps between the documented inputs and outputs.

FDA 21 CFR Requirements – Design Output

21 CFR Part 820.30(D) Design Output per FDA

- *"Each manufacturer shall establish and maintain procedures for defining and documenting design output in terms that allow an adequate evaluation of conformance to design input requirements.*
- *Design output procedures shall contain or make reference to acceptance criteria and shall ensure that those design outputs that are essential for the proper functioning of the device are identified.*
- *Design output shall be documented, reviewed, and approved before release.*
- *The approval, including the date and signature of the individual(s) approving the output, shall be documented."*

During this phase, product specifications (PS) and the device master record (DMR) are generated to define the design output. Planning for process validation and manufacturing begins during this phase often with the creation of a validation master plan (VMP). In any design office or factory setting, a lot of data and paperwork are generated. Therefore, it is important to be able to make the distinction between what is a design output and what is not. The first way of identifying a design output is to verify if it is listed as a task, a deliverable or listed in the design and development plan. If this is the case, then it is classified as a design output. Furthermore, if it describes or defines a design feature, it can also be classed as a design output.

The quality system requirements for design output can be separated into two elements: Design output should be expressed in terms that allow adequate assessment of conformance to design input requirements and should identify the characteristics of the design that are crucial to the safety and proper functioning of the device. This raises two fundamental issues for developers:

What constitutes design output?

The first issue is important because the typical development project produces voluminous records, some of which may not be categorized as design output. On the other hand, design output must be reasonably comprehensive to be effective. As a general rule, an item is a design output if it is a work product, or deliverable item, of a design task listed in the design and development plan, and the item defines, describes, or elaborates an element of the design implementation. Examples include block diagrams (software coding), flow charts, software high-level code, and system or subsystem design specifications. The design output in one stage is often part of the design input in subsequent stages. Design output(s) includes production specifications as well as descriptive materials which define and characterize the design. As a project develops the production requirements can be added to a Process control plan.

Production Specifications

Production specifications draw upon many documents that are used to manufacture, test, inspect, install, maintain and service a device. They include: (1) component and material specifications, (2) production and process specifications, (3) work instructions and SOPs, (4) quality plans, specifications and procedures, (4) labelling specifications, and (5) packaging specifications.

Design Review

Formal design reviews are critical to the efficacy of design control, and ultimately, the market success of the device. They should be planned for up front in the design development plan. Changes late in the design cycle are much more expensive than those made early on. Design reviews can play an important role in identifying changes in a timely manner and thus prevent costly redesigns close to the launch date. The FDA QSR clearly specifies the need for independent reviewers. Independent reviewers must be far enough removed from the design in order to provide an objective review.

FDA 21 CFR Requirements- Design Review

FDA CFR Part 820.30(E) Design review per FDA

- *"Each manufacturer (of medical devices) shall establish and maintain procedures to ensure that formal documented review(s) of the design results are planned and conducted at phases of the device's design development.*
- *The procedure(s) shall ensure that participants at each design review include representatives of all functions (e.g. engineer, quality, manufacturing etc) concerned with the design phase being reviewed and an individual(s) who does not have direct responsibility for the design phase being reviewed, as well as any specialists needed.*
- *The results of a design review, including identification of the design, the date, and the individual(s) performing the review, evidence and minutes of same shall be documented in the design history file (the DHF)."*

Key goals of design review:

- provide feedback to designers on existing or emerging problems
- assess project progress
- provide confirmation that the project is ready to move on to the next phase of development

Many types of reviews occur during the course of developing a product. Reviews may have both an internal and external focus. However, each stage of design control must have a review where the progress is reviewed by a core team and independent reviewer.

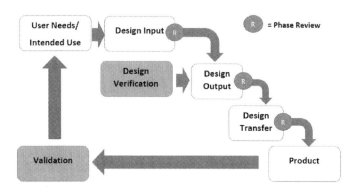

Reviews are important in ensuring that the input requirements are not forgotten as the project progresses. Secondly, there must be "agreement" between the user requirements and design inputs versus the design outputs.

A formal review of the design input requirements early in the development process is normally completed. The number of reviews depends upon the complexity of the device.

For a simple design or product, or a minor upgrade to an existing product, it might be appropriate to conduct a single review at the conclusion of the design process.

For a product involving multiple subsystems, an early design task is to allocate the design input requirements among the various subsystems. For example, in a microprocessor-based system, designers must decide which functions will be performed by hardware and which by software. In another case, tolerance build-up from several components may combine to create a clearance problem.

System designers must establish tolerance specifications for each component to meet the overall dimensional specification. In cases like these, a formal design review is a prudent step to ensure that all such system-level requirements have been allocated satisfactorily prior to engaging in detailed design of each subsystem.

Reviewers

In determining who should participate in a formal design review, planners should consider the qualifications of reviewers and the types of expertise required. Often a matrix listing reviewers, functions and responsibilities is documented in a design control procedure or at the outset of a project.

Evaluation of the design

Many formal design reviews take the form of a meeting. At this meeting, the designer(s) may make presentations to explain the design implementation, and persons responsible for verification activities may present their findings to the reviewers. Reviewers may ask for clarification or additional information on any topic, and add their concerns to any raised by the presenters. This portion of the review is focused on finding problems, not resolving them. There are many approaches to conducting design review meetings. In simple cases, the technical assessor and reviewer may be the same person, often a project manager or engineering supervisor, and the review meeting is a simple affair in the manager's office. For more elaborate reviews, detailed written procedures are desirable to ensure that all pertinent topics are discussed, conclusions accurately recorded, and action items documented and tracked.

There is a dangerous tendency for design review meetings to become adversarial affairs. The reputation of the designers tends to be linked to the number of discrepancies found, causing the designers to become defensive, while the reviewers score points by finding weaknesses in the design.
The resulting contest can be counterproductive. An added complication is the presence of invited guests, often clinicians, who are expected to provide the user perspective. These reviewers are often very reluctant to ask probing questions, especially if they sense that they may become involved in a conflict where all the rules and relationships are not evident.

These difficulties can be avoided by stating the goals and for conducting the formal design review clearly at the outset. While the designers are in the best position to explain the best features of the design, they are also most likely to be aware of the design's weaknesses. If the designers and reviewers are encouraged to work together to systematically explore problems and find solutions, the resultant design will be improved and all parties will benefit from the process.

Each function and area of expertise should be encouraged to ask questions, avoid making assumptions, and think in a critical fashion.

For extremely simple designs or design changes, it may be appropriate to specify a procedure in which a review is completed and approved by the required parties and independent review.

Implementation of corrective actions

Not all identified concerns result in corrective actions. The reviewers may decide that the issue is not of concern or immaterial. In most cases, however, resolution involves a design change, a requirements change, or a combination of the two. If the fix or solution is evident, the reviewers may specify the appropriate corrective action. Alternatively, a task may be taken by a team member to complete concurrently.

In any case, action items and corrective actions are normally tracked under the manufacturer's change control procedures.

Design Inputs-Outputs and Verification (Matrix), IOV

The IOV matrix is a document that ensures all inputs are documented and associated with a design output and acceptance criteria. An example of a specific design input would include a product requiring sterility. The input requirement would be sterilisation (e.g. autoclaving). The design output along with the acceptance criteria would be "the product is sterile to a Sterility assurance level of 1×10^{-6}.

Next is the Design Verification Documentation or Design Validation documentation (as applicable) which should detail the report or document number related to the "sterility" testing.

Relationship Of Design Review To Verification And Validation

In practice, design review, verification, and validation overlap one another, and the relationship among them may be confusing. As a general rule, the sequence is: verification, review, validation, review. In most cases, verification activities are completed prior to the design review, and the verification results are submitted to the reviewers along with the other design output to be reviewed. Alternatively, some verification activities may be treated as components of the design review, particularly if the verification activity is complex and requires multidisciplinary review. Similarly, validation typically involves a variety of activities, including a determination that the appropriate verifications and reviews have been completed. Thus, at the conclusion of the validation effort, a review is usually warranted to assure that the validation is complete and adequate.

Both Clinical and non-clinical testing may be required to form the basis of verification activities when completed to determine what the design output specifications are. Typically, verification activities involve some scientific or chemical analysis and establish that the output specification is correct. Verification activities must be appropriate for the particular output that is being verified and documented.

Verification activities are completed via means of chemical/microbial analysis, inspection, measurement of an attribute, analysis or testing via other benchtop equipment. Three categories of verification include:

Standardised verification
Non-Standardised
Creative verification

Standardised verification are conducted uner controlled conditions and that meet a set of requirements established in industry industry. They are typically documented in recognised standards such as ISO, ANSI, USP, etc. These methods have already been validated to prove some aspect of design and no validation by the designer is needed, only proof that the standard was followed.

Non-standardised verification activities include a variety of tests, inspections, or analyses that can vary from product to product. These may be recognised methods that are described in literature or scientific publications and constitute well-understood principles but are not described in a recognised standard. The designer will need to validate that the set-up and application of these methods are applicable to their design and will provide the necessary results.

Creative verification activities require the manufacturer to be creative in devising a plan or coming up with activities that can be used to verify a particular aspect of design.

When there are no known standardised methods or recognised non-standard methods to prove an aspect of design, the manufacturer will need to devise their own test, inspection or analytical method. These methods must be validated to show they will provide the required results and that the results will be applicable to the design.

Documenting Verification

The methods, plans, and protocols used for verifying design must be documented, reviewed and approved. All documents that provide evidence in support if verification results need to be added to the DHF.

Design Verification, Validation and Transfer Phase

To illustrate the concepts, consider a building design. In a typical scenario, the senior architect establishes the design input requirements and sketches the general appearance and construction of the building, but contractors typically elaborate and interpret the details into practical terms. Verification refers to the checking at each phase to ensure the output meets the design requirements. For example, if a device is designed to take both AC electrical power and a battery (DC power), the design engineer must verify that these are accounted for in the plans and production specifications.

FDA 21 CFR Requirements - Design Verification

FDA CFR Part 820.30(f) Design Verification

- *"Each manufacturer shall establish and maintain procedures for verifying the device design.*
- *Design verification shall confirm that the design output meets the design input requirement(s).*
- *The results of the design verification, including identification of the design, method(s), the date, and the individual(s) performing the verification, shall be documented in the Design History File."*

The ultimate aim of design verification is to finalise design specification. Examples of verification activities include:

- Design failure modes and effects analysis (DFMEA)
- Fault tree analysis
- Package integrity tests
- Biocompatibility testing
- Bioburden testing of packed products
- Worst case analysis – tolerance stacking of components

Design Validation

Design validation of the product is nesessary to ensure the device meets the user requirements and intended use. Above all, it ensures the device operates reliably and safely. Process validation is required in order to confirm manufacturing specifications and the Device Master Record (DMR).

Planning Validation

FDA regulations require device manufacturers to establish and maintain procedures for validating the device design.

A plan for validation during the initial stages of the design process should be developed and documented. The validation methods, procedures, and acceptance criteria should be established, documented, and maintained.

Validation is established via the documented evidence that the device as designed will perform and function as defined in the initial design input requirements and will meet defined intended uses as well as the needs of users/customer in the actual use environment.

Validation Methods

Design validation must be performed under an actual or simulated use conditions with an established protocol that is preapproved.

If design verification provides data to the design engineer relative to the appropriateness, correctness, and robustness of their design solution, then design validation provides evidence to users that the chosen design solution will perform and function in the manner described on the label, and labelling and will meet the stated intended uses as well as user needs- and above all is safe.

Like design verification, design validation can consist of inspection, measurement, analysis or test. Unlike design verification, most design validation activities will be tests. Most clinical trials or studies are done as part of design validation to prove the performance and functionality as well as how well the needs of the users were actually met. When clinical trials or studies are performed, they are done in the actual use environment. User surveys, usability studies, and other types of user interactions may also be design validation activities if they are used to answer user needs issues, including human factors.

Design validation can be completed using actual or simulated conditions of us. Therefore, bench or laboratory testing may be sufficient for Design validation purposes.

Design validation addresses:

- Intended use
- Needs of patient/user
- Packaging and labelling
- Environmental simulation

The manufacturer must ensure that validation proves their device design performs and functions as described in their product labelling and actually meets their intended use described via a URS.

Validation must also address product packaging and labelling. These areas of design may have significant human factors implications and could affect products in unexpected ways (e.g. misuse).

Validation must include simulation of the product in environmental conditions as close as possible to those it will be exposed to, whether during shipping, handling, storage, or use.

DESIGN VALIDATION TYPICALLY INVOVLVES CONDUCTING IN VIVO TESTING CLINICAL TRIALS OR CLINICAL EVALUATION.

FDA 21 CFR Requirements - Design Validation

FDA CFR 820.30(G) Design Validation as per FDA

- *"Each manufacturer shall establish and maintain procedures for validating the device design.*
- *Design validation shall be performed under defined operating conditions on initial production units, lots, batches, or their equivalents.*

- *Design validation shall ensure that devices conform to defined user needs, intended uses and shall include testing of production units under actual or simulated use conditions.*
- *Design validation shall include software validation and risk analysis, where appropriate.*
- *The results of the design validation, including identification of the design, method(s), the date, and the individual(s) performing the validation, shall be documented in the design history file."*

Verification examines design outputs at the different phases of the process while design validation confirms that all user needs are achieved even when subject to anticipated sources of variation such as materials, processing equipment, suppliers and so on.

Validation Review

Validation may expose deficiencies in the original assumptions concerning user needs and intended uses. A formal review process should be used to resolve any such deficiencies. As with verification, the perception of a deficiency might be judged insignificant or erroneous, or a corrective action may be required.

Validation Methods

Many medical devices do not require clinical trials. However, all devices require clinical evaluation and should be tested in the actual or simulated-use environment and conditions as a part of validation. This testing should involve devices which are manufactured using the same methods and procedures expected to be used for ongoing production.

While testing is always a part of validation, additional validation methods are often used in conjunction with testing, including analysis and inspection methods, compilation of relevant scientific literature, provision of historical evidence that similar designs And/or materials are clinically safe, and full clinical investigations or clinical trials.

Some manufacturers have used their best workers or technicians to produce products or materials for a trial or evaluation. However, this should be cautioned as it may not take into account variation within the manufacturing process.

The FDA advises that pilot production should simulate as closely as possible the actual manufacturing conditions.

Validation should also address product packaging and labelling. These components of the design may have significant human factors that may affect product performance in unexpected ways. For example, packaging materials have been known to cause electrostatic discharge (ESD) failures in electronic devices. If the unit under test is delivered to the test site in the test engineer's briefcase, the packaging problem may not become evident until after release to market.

Validation should include simulation of the expected environmental conditions. Core inputs to environmental testing includes temperature, humidity, shock and vibration (shock and vibe), etc. It should be noted that for some classes of device, the environmental stresses encountered during shipment and installation far exceed those encountered during actual use and should be addressed during validation.

Design Transfer

The purpose of design transfer is to finalise all deliverables for filing with regulatory agencies.

FDA 21 CFR Requirements - Design Transfer per FDA

FDA CFR Part 820.30(H) Design Transfer

- *Each manufacturer shall establish and maintain procedures to ensure that the device design is correctly translated into production specifications.*

The aim of design control is to eventually provide a design of a product that can be transferred into production specifications (drawings, manufacturing, test, and inspection procedures). Production specification(s) must ensure that manufactured devices are **consistently and reliably** produced within product and process capabilities, meeting all quality requirements. Therefore, consistency and reliability of equipment and processes should be understood and challenged accordingly. No design team can anticipate all factors bearing on the success of the design, but procedures for design transfer should address at least the following basic elements.

Firstly, the design and development procedures should include a qualitative assessment of the completeness and adequacy of the production specification- is there adequate information of material type, dimensional tolerances etc.

Second, the procedures should ensure that all documents which constitute the production specifications are reviewed and approved.

Third, the procedures should ensure that only approved specifications are used to manufacture production devices.

Post-Launch Reviews

A post-launch review is required for each product within one year of initial launch. The purpose of the post-launch review is to confirm that no design or manufacturing changes are required and to document future product development activity. It also considers performance and patient safety. A mechanism should be established to track all change requests and change orders to ensure proper disposition.

Post-transfer Design Changes

Design change control continues throughout the product life and is managed in accordance with the change management process which exists within a company. Design changes including labeling updates, material changes etc should be assessed for impact to design, most importantly design verification and/or design validation. New studies or engineering activities may be required to support any changes and then provide the updated Verification and validation evidence.

If post transfer design changes occur, the DHFI will be updated annually with design change information. This information should include the design change request number (ie: QCR number) and design change title/description. Post transfer updates to the DHFI shall be approved by the appropriate Global Quality member.

Common Design Control Deliverables

This section provides a non-exhaustive list of design documentation deliverables. A brief description of each is provided. This list can be used as a checklist for the design control process or as supplementary information of key activities outlined previously.

Validation Master Plan (VMP): A validation master plan should be written as soon as the project begins. It should describe the product to be manufactured and the process technology. A VMP will also contain generic material such as an outline of the validation approach and the types of validation e.g. prospective, concurrent and so on.

External Requirements: External requirements refer to regulations and industry standards that are relevant to a new product. At the design input phase a list of documents should be created in order to capture essential requirements as early as possible.

Design Development Plan: A design and development plan is an overarching document that describes the design and development, responsibilities, timelines and project scope, list and schedule of major tasks and the phase review details such as the timing and approval requirements.

Product Specification: The product specification is a design output document that is built over the course of the project. Not all information will be final in the early phases, however, having an early draft will help focus minds and generate the right activity in order to define target dimensions, physical attributes and tolerances.

Stability Testing: A document containing a summary of results, testing and analysis should be created and filed as part of the DHF.

Design Inputs-Outputs and Verification (Matrix), IOV

The IOV template is a document that ensures the design meets the user needs and they are detailed as Design Input(s), Output(s), and Verification and Validation is completed as applicable, with references to reports/studies provided.

Device Master Record: A DMR is an output document and should be available at the design transfer phase. It is a comprehensive list referencing all work instructions, test procedures, test specifications, manufacturing specifications and finished product specifications required to manufacture the product.

Test Method Validations: A list of all validated test methods (functional, analytical, physical etc.) should be available to file in the DHF.

Design History File: The DHF is a repository for all of the documentation generated as a result of the design control process. The DHF serves as a complete record of the design.

Design Control Process via Web-Based Systems

In recent years some companies have entered the market offering web-based design control processes. As mentioned earlier, there are a large amount of documents created during the design control process. Most of the documentation generated is subject to change control and therefore requires review and approval. As with traditional hardcopy approval, this can be time-consuming and complex if approvers are based across different departments or drawn internationally. All documents also form part of the design history file. Therefore, the proper filing and availability of documents is an important source of concern. The use of an electronic system may mitigate some of these concerns. Furthermore, some web-based solutions offer integration with existing electronic documentation systems or integration

Risk Assessments for Design Controls

Severity: A measure of the possible consequences of a hazard.

Probability of Failure: An estimate of the likelihood of failure.

Detection of Failure: An estimate of the likelihood of detection.

Risk Priority Number (RPN): The product of ratings on occurrence, severity and detection.

CAPA: Corrective and Preventative Action.

ALARP: As Low As is Reasonably Practicable.

Verification Plan: A confirmation through the provision of objective evidence, that the specified requirements have been fulfilled.

Failure Modes and Effects Analysis, FMEA: A formal risk assessment methodology/tool for identifying potential failure modes, and assigning numerical values to the severity, likelihood of occurrence, and likelihood of escaped detection to failure modes in order to quantify risk.

Risk Analysis
Risk analysis can be performed using a variety of methodologies such as FMEA/FMECA, HAZOP.

Review/Inspection of Design Controls

1 Selection and identification of a single design project-
Verify that design control procedures are in place that address the requirements of Section FDA 21 CFR Part 820.30 of the regulation have been defined & documented.

2 Review the design-plan for the project to understand the layout of the design and development activities and the assigned responsibilities and interfaces.

3 Confirm that design inputs were established for the project/product.

4 Verify that design outputs essential for the proper functioning of the device were identified and documented.

5 Confirm that acceptance criteria were established in advance of the performance of verification activities and validation activities.

6 Determine if design verification confirmed that design outputs met the specified design input requirements for the product.

7 Confirm that design validation data show that the approved design met the predetermined user needs and intended uses.

8 Confirm that the completed design validation was completed without any discrepancies.

9 If the device uses software, confirm that the software was Validated in accordance with FDA requirements.

10 Confirm that risk analysis was performed as required.

11 Determine if design validation was accomplished using initial production devices or equivalent devices.

12 Confirm that changes were controlled including validation or where appropriate verification

13 Determine if design reviews were conducted.(per procedure)

14 Determine if the design was correctly transferred. (Design Transfer)

Inspection and Review in detail ...

1. FDA CFR Section 820.30 Design controls apply to the design of Class II and III medical devices, and a select group of Class I devices. As previously stated, the regulation allows for medical device companies to establish design controls that make best-sense for their sub-industry and the particular product type.

Software validation is included in FDA 21 CFR Section 820.30(g) Design Validation. If a device employs software, a review of the software's validation is also prudent while completing the assessment of the design control management system.

2. Both Written and electronic procedures are required by companies manufacturing medical devices. Procedures form the basis of the design control system and must be maintained and reviewed periodically as defined by the company. An inspection of Design control procedures should determine if the design input procedures includes a mechanism for addressing incomplete, unclear or conflicting requirements.

Design output SOPs and procedures ensure that critical design outputs for the functioning and operation of the device are identified and the design review procedure ensures that each design review.

Even within the research and development stage, some level of change control, management oversight and Good Engineering Practice (GEP) should be applied. After the project design is chosen, and a decision is made to develop a particular design, a design plan must be established.

Design control can begin at this point, but *must* begin prior to the first set of design inputs been established. Plans must define responsibility for implementation of the design and development activities and identify and describe the relationships with different groups or activities.

Verify that the design outputs that are essential for the proper functioning of the device were identified.

Inputs are the requirements of a device based on the intended use. Sources of design inputs include the voice of the consumer, regulations, standards (ISO, ASTM) etc. Review the sources used to develop inputs. The design inputs should cover all relevant aspects such as *intended use, risk, performance characteristics, biocompatibility, compatibility, human factors, sterility* and the intended environment of use.

The FDA recommends that "Outputs must be comprehensive enough to characterize the device design to allow for verification and validation."

Design outputs which are essential for the proper functioning of the device must be identified. Typically a risk analysis tool such as an FMEA is used to determine essential outputs.

Verification and validation activities must include Acceptance criteria documented in a protocol prior to testing. Review the documentation associated with a sample of verification activities and a sample of validation activities as determined using the Sampling Tables. If possible, select activities that are associated with outputs that address critical functioning. Confirm that acceptance criteria were established prior to performance of the verification or validation.

Design verification activities are performed to provide objective evidence that design output meets the design input requirements. Verification activities can include tests, inspections, analyses, measurements, and so on.

Design validation is performed to gain objective evidence that device specifications (outputs) conform with user needs and intended use(s). Design validation must be completed before commercial distribution and sale of the device.

Design validation involves the performance of clinical evaluations and includes testing under actual or simulated use conditions. Clinical evaluations can include clinical investigations or clinical trials, but they may only involve other activities. These may include evaluations in clinical or non-clinical settings, provision of historical evidence that similar designs are clinically-safe, or literature reviews.

Validation activities must address the needs of all relevant parties (i.e. patient, health care worker, etc.) and be performed for each intended use. Validation activities should address the design outputs of labeling and packaging.

Design validation may detect discrepancies between the device specifications (outputs) and the needs of the user or intended use(s) of the device. All discrepancies must be addressed and resolved by the firm. This can be accomplished through a change in design output or a change in user need or intended use.

design validation also includes the requirement for software validation. If the selected device is software controlled its software must be validated. Process validation may be conducted concurrently with design validation. Production devices used in design validation may have been manufactured in a production run during process validation.

The Quality System regulation clarified and relocated the requirement into Section 820.30(i). It expanded the requirement to include changes made during the design process (pre-production changes).

Change control is not a new requirement. The 1978 GMP regulation Section 820.100(a)(2) required approval of changes made to specifications after final design transfer (post-production changes).

The documentation and control of design changes begin when the initial design inputs are approved and continues for the life of the product. Examples of the application of change control include: changes made to approved inputs or outputs such as to correct design deficiencies identified in the verification and validation activities; labeling changes; changes which enhance the device's capabilities or the capabilities of the process; and changes resulting from customer complaints.

Product development is inherently an evolutionary process. While change is a healthy and necessary part of product development, quality can be ensured only if change is controlled and documented in the development process, as well as in the production process.

Post-production design changes require the firm to loop back into the design controls of Section 820.30 of the regulation. This does not mean that post-production changes have to go back to the R&D Department for processing. This track is dependent on what the firm specifies in their change procedure. It is acceptable for the manufacturing department to process the entire design change and to implement the controls of Section 820.30.The design change control section is linked to and is redundant with Section 820.70(b) Production and process changes of the regulation.

All design changes must be verified and documented. Design changes must also be validated unless the performance of only verification can be justified and documented.

When a design change cannot be verified by inspection or test, it must be validated.

Formal design reviews are planned and typically conducted at the end of each design stage or phase, or after completion of project milestones. The number of reviews is dependent on the complexity of the design. A single review may be appropriate at the conclusion of the design project for a simple design or a minor change to an existing product. Multiple reviews are typically conducted for projects involving subsystems or complex designs.

Feedback should be gained by designers from design reviews on existing or emerging problems. The design review process should account for risk analysis and change control where relevant.

Medical Device Software and "Apps"

Software applications also known as "apps" have become a mainstream phenomenon for mobile phones and tablets and other personal electronic devices. Apps have crossed over into the health and fitness sector that allow personal devices to track and report physical activity, sleep and provide health tips. Furthermore, medical devices also offer increased interaction with "apps" allowing real-time analysis of physiological data.

How to determine if a software app is a medical device?

In simple terms, any software app that provides a diagnostic function in order to determine disease or a medical condition is likely to fall under the definition of a medical device. Examples include:

- Apps that calculate medicine doses a patient is to take
- Apps that identify or inform you that you have a particular medical condition or disease
- Apps provide a risk-based assessment

CE marking of Devices

When an app developer applies a 'CE mark' they are claiming that the app is fit for the purpose (aka the intended purpose) and it is acceptably safe to use. The CE mark should be visible on the app when you are looking at it in the app store, the 'landing' page or on the app developer's product website or information page. The manufacturer has a duty to provide clear information that describes what the app can be used for and how to use it. Consumers should exercise caution when purchasing apps from unknown sources. Product not CE marked or products not assessed for safety pose a potential risk when used. Any medical device app that does not have a CE mark evident from literature, manufacturers information or interface should be reported to the relevant competent authority.

Personal data and apps

Many apps have the ability to capture and record physiological parameters such as heart rate, sleep patterns and activity. Many also support the input of additional data such as a persons weight and physical attributes.

It is very important that you have read the small print to understand what personal data you may have agreed to share with the developer by signing up to the app and how they might store or use your data or share your information with third parties. This includes information about you such as your name, address, date of birth and information about your health.

General Guidance on Use

Once you are sure the app is right for you and it is CE marked then you should follow the instructions carefully.

Be honest with the information you put into the app. If you enter wrong information about yourself, the app may not give you the right result. Ensure that you always update the app to the newest compatible version. After using
If you are in doubt about the information that the app has given you or you are concerned about your health then you should consult a healthcare professional (a pharmacist, health visitor, practice nurse or GP) If you have any problems with the app not working as stated e.g.

• If the instructions aren't clear or the app is difficult to use
• If the app isn't giving you the results that you expected
• If you have concerns over the safety of the app or the information that it provides

European Requirements –Stand-alone Software

This guideline on the qualification and classification of stand-alone software was drafted by the European Commission after consultation of the competent authorities, commission services, industry and notified bodies

MEDDEV 2.1/6 rev. 1: Qualification and classification of stand-alone software

A stand-alone software must meet the following criteria in order to be classified as a medical device:

- o it has to be a computer program

- o the software has to have a different purpose than mere storage, archival, lossless compression, communication or simple search

- o the software has to be for the benefit of individual patients

- o the software has to have an intended purpose listed in Article 1(2)a) of Directive 93/42/EEC.
  (In Germany, Article 1(2)a) of Directive 93/42/EEC has been implemented as national law in Section 3 number 1 MPG.)

FDA Requirements -Medical Device Software

Food and Drug Administration FDA: "Mobile Medical Applications - Guidance for Industry and Food and Drug Administration Staff", February 2015

Introduction

Issued in 2015, the FDA provided non-binding guidance on "Mobile Medical Applications". The aim of the guidance is to inform industry on how the FDA intends to apply its regulatory authorities to select software applications intended for use on mobile platform/mobile apps. The guidance also states that although some mobile apps may meet the definition of a medical device, if they propose a lower risk to the public FDA intends to exercise enforcement discretion over these devices.

As mobile platforms become more user friendly, computationally powerful, and readily available, innovators have begun to develop mobile apps of increasing complexity to leverage the portability mobile platforms can offer. Some of these new mobile apps are specifically targeted to assisting individuals in their own health and wellness management. Other mobile apps are targeted to healthcare providers as tools to improve and facilitate the delivery of patient care.These software devices include products that feature one or more software components, parts, or accessories (such as electrocardiographic (ECG) systems used to monitor cardiac rhythms), as well as devices that are composed solely of software (such as laboratory information management systems).

On February 15, 2011, the FDA issued a regulation down classifying certain computer- or software-based devices intended to be used for the electronic transfer, storage, display, and/or format conversion of medical device data – called Medical Device Data Systems (MDDSs) – from Class III (high-risk) to Class I (low-risk).2 The FDA has previously clarified that when stand-alone software is used to analyze medical device data, it has traditionally been regulated as an accessory to a medical device3 or as medical device software.

As is the case with traditional medical devices, certain mobile medical apps can pose potential risks to public health. Moreover, certain mobile medical apps may pose risks that are unique to the characteristics of the platform on which the mobile medical app is run interpretation of radiological images on a mobile device could be adversely affected by the smaller screen size, lower contrast ratio, and uncontrolled ambient light of the mobile platform. FDA intends to take these risks into account in assessing the appropriate regulatory oversight for these products. This guidance clarifies and outlines the FDA's current thinking.

Definitions

Mobile Platform: For purposes of this guidance, "mobile platforms" are defined as commercial off-the-shelf (COTS) computing platforms, with or without wireless connectivity, that are handheld in nature. Examples of these mobile platforms include mobile computers such as smart phones, tablet computers, or other portable computers.

Mobile Application (Mobile App): For purposes of this guidance, a mobile application or "mobile app" is defined as a software application that can be executed (run) on a mobile platform (i.e., a handheld commercial off-the shelf computing platform, with or without wireless connectivity), or a web-based software application that is tailored to a mobile platform but is executed on a server.

Mobile Medical Application (Mobile Medical App): For purposes of this guidance, a "mobile medical app" is a mobile app that meets the definition of device in section 201(h) of the Federal Food, Drug, and Cosmetic Act (FD&C Act) - 7 - 4 ; and either is intended:

to be used as an accessory to a regulated medical device; or
to transform a mobile platform into a regulated medical device.

Mobile Medical App: the FDA defines a "mobile medical app manufacturer" is any person or entity that manufactures mobile medical apps in accordance with the definitions of manufacturer in 21 CFR Parts 803, 806, 807, and 820. They may include companies that initiates specifications, designs, labels, or creates a software system or application for a regulated medical device in whole or from multiple software components.

BfArM Germany

Guidance from the Federal Institute for Drugs and Medical Devices (BfArM), Germany

Differentiation between apps and medical or other devices as well as on the subsequent risk classification according to the MPG

BfArM as with other competent Authorities provides guidance on differentiation between:

(1) apps (in general: stand-alone software, not incorporated into a medical device, e.g. as control software) and

(2) medical devices.

In similar fashion, any this guidance is informative and non- binding. Qualification and classification needs to determine the intended purpose of the software and its classification must is the responsibility of the manufacturer.

- Differentiation/Qualification
- Risk Classification
- Examples for Differientation/qualification
- Further Guidance
- European Commission
- Other Authorities
- Committees

The "intended purpose" is the use for which the medical device is intended according to the manufacturer's information, marketing, labelling and instructions for use (IFU).

Thus, not only the explicitly described intended purpose is relevant e.g. for an authority decision on qualification as a medical device, but also the instructions for use and the promotional materials (e.g. website, information in App Store) regarding the specific product.

BfArM guidance also points out that Stand alone software such as smartphone apps can indeed be classified as a medical device. However, the product must be intended by the manufacturer to be used for humans with a minimum of at least one of the criteria below fufilled:

- diagnosis, prevention, monitoring, treatment or alleviation of disease,

- diagnosis, monitoring, treatment, alleviation or compensation of injuries or handicaps,

- investigation, replacement or modification of the anatomy or of a physiological process,

- control of conception.

Essentially, the above criteria is a summary of the European regulations pertaining to medical devices. As opposed to mere provision of knowledge, e.g. in a paper or electronic book (no medical device), any type of interference with data or information by the stand-alone software is indicative of a classification as a medical device. Possible indicative terms in connection with the intended purpose of corresponding functions can be e.g.: alarm, analyse, calculate, detect, diagnose, interpret, convert, measure, control, monitor, amplify. Indicative functions for classification as a medical device can be among the following:

- decision support or decision-making software e.g. with regard to therapeutic measures

- calculation e.g. of dosing of medicines (as opposed to mere reproduction of a table from which users can deduce the dosage themselves)

- monitoring patients and collecting data e.g. by measurements if the results thereof have an influence on diagnosis or therapy.

Pure data storage, archiving, lossless compression, communication or simple search functions do not result in classification as a medical device. Like all other medical devices from own production, software applications from own production are medical devices and thus must fulfil the basic requirements of Council Directive 93/42/EEC.

Risk Classification

With the exception of in vitro diagnostic medical devices and active implantable medical devices, medical devices are allocated to risk classes that are mainly based on the potential damage that can be caused by an error/malfunction of the medical device. These risk classes range from Class I (low risk) and IIa and IIb to Class III (high risk). Class I products are additionally subdivided according to whether they require sterilisation (Is) or include a measuring function (Im) which is relevant for the further conformity assessment procedure. The classification is based on the rules laid down in Annex IX of Council Directive 93/42/EEC. The following rules are most suitable for the classification of stand-alone software.

Rule 9

"All active therapeutic devices intended to administer or exchange energy are in Class IIa unless their characteristics are such that they may administer or exchange energy to or from the human body in a potentially hazardous way, taking account of the nature, the density and site of application of the energy, in which case they are in Class IIb. All active devices intended to control or monitor the performance of active therapeutic devices in Class IIb, or intended directly to influence the performance of such devices are in Class IIb."

Rule 10

"Active devices intended for diagnosis are in Class IIa,

- if they are intended to supply energy which will be absorbed by the human body, except for devices used to illuminate the patient's body, in the visible spectrum;

- if they are intended to image in vivo distribution of radiopharmaceuticals;

- if they are intended to allow direct diagnosis or monitoring of vital physiological processes, unless they are specifically intended for monitoring of vital physiological parameters, where the nature of variations is such that it could result in immediate danger to the patient, for instance variations in cardiac performance, respiration, activity of CNS in which case they are in Class IIb.

- Active devices intended to emit ionizing radiation and intended for diagnostic and therapeutic interventional radiology including devices which control or monitor such devices, or which directly influence their performance, are in Class IIb."

Rule 12

"All other active devices are in Class I."

Rule 14

"All devices used for contraception or the prevention of the transmission of sexually transmitted diseases are in Class IIb, ..."

The following definitions in accordance with Annex IX Section I No. 1 of Council Directive 93/42/EEC are to be observed:

- **Stand alone software**

  Stand alone software is considered to be an active medical device.

- **Active therapeutic device**

  "Any active medical device, whether used alone or in combination with other medical devices, to support, modify, replace or restore biological functions or structures with a view to treatment or alleviation of an illness, injury or handicap."

- **Active device for diagnosis**

  "Any active medical device, whether used alone or in combination with other medical devices, to supply information for detecting, diagnosing, monitoring or treating physiological conditions, states of health, illnesses or congenital deformities."

The afore-mentioned rules show that e.g. medical apps on smartphones and tablets will mostly be classified in risk Class I in accordance with Rule 12. If the medical devices are intended for diagnosis or monitoring of vital functions (e.g. cardiac functions), Classes IIa or IIb can also be considered.

Depending on the risk class there are different requirements for conducting a conformity assessment procedure as the prerequisite for affixing the CE marking and for correct marketing within the European Economic Area.

Thus, the manufacturer can perform a conformity assessment e.g. for Class I devices without involvement of a notified body; for all other risk classes (also in the case of Class I devices that require sterilisation or include a measuring function) it is mandatory to involve a notified body. If a stand alone software or app is placed on the market as a medical device it is subject to the same regulations as all other medical devices.

Examples for Qualification/differentiation

Decision supporting software

In general, software is usually considered a medical device when it is used for healthcare, if e.g. medical knowledge databases and algorithms are combined with patient-specific data and the software is intended to give healthcare professionals recommendations on diagnosis, prognosis, monitoring or treatment of an individual patient.

Software systems

If a software consists of several modules it is the manufacturer's responsibility whether he wants the modules as a whole to be qualified and classified or each module individually. If the entire system is qualified and if it consists both of software with and without the properties of a medical device, the system is subject to medical device legislation.

Telemedical software

In telemedicine the physician observes and assesses the patients' medical data using telecommunication technologies - e.g. via the internet. Patient and physician can be at different locations. Depending on the intended purpose, communication systems for telemedicine can either be non-medical devices (purely for transfer of data) or a combination of non-medical devices and medical devices (e.g. in order to support diagnoses).

Hospital information systems (HIS)

Hospital information systems that support patient management are generally not medical devices, especially if they have the following intended purpose:

- o collection of data for patient admission
- o administration of general patient data
- o scheduling of appointments

o   insurance and billing functions

However, hospital information systems can be combined with other modules that could be medical devices.

Picture Archiving and Communication System (PACS)

For example, if the manufacturer of the PACS software specifies in the intended purpose that the software is only meant for storage or archiving of pictures and not for diagnosing, this would indicate that it is not a medical device. However, if the manufacturer intends the PACS software for controlling a medical device or to have an influence on its use or to allow a direct diagnosis, this would support its classification as a medical device.

Stand alone software or apps that are **not** medical devices

Operating system software:
Operating system software (e.g. Windows, Linux) is neither a medical device nor is it an accessory to a medical device.

Software for general purposes

Software for general purposes is not a medical device even if it is used in connection with healthcare.

Software or apps as health or fitness products
When differentiating medical devices e.g. from health or fitness products, the decisive issue is whether they are intended for medical or non-medical purposes. This is defined by the manufacturer of the product. Software or apps merely intended for sporting activities, fitness, well-being or nutritional aims without a medical purpose claimed by the manufacturer are generally not medical devices

Medicines and Healthcare products Regulatory Agency, MHRA

The Medicines and Healthcare products Regulatory Agency (MHRA) is an agency of the Department of Health in the UK which is responsible for ensuring that medicines and medical devices work and are acceptably safe.

Formed in 2003 with the merger of the Medicines Control Agency (MCA) and the Medical Devices Agency (MDA). In April 2013, it merged with the National Institute for Biological Standards and Control (NIBSC) and as the MHRA. The MHRA has released guidance on Medical Device stand-alone software including apps (Aug 2014). This guidance should be used in addition to MEDDEV 2.1/6.

Medicines & Healthcare products Regulatory Agency (MHRA): Guidance - Medical device stand-alone software including Apps, August 2014

Medical device stand-alone software including apps (including IVDMDs)

As well as medical device apps becoming a growth area in healthcare management in hospital and in the community settings, the role of apps used as part of fitness regimes and for social care situations is also expanding. However, in the UK and throughout Europe, standalone software and apps that meet the definition of a medical device are still required to be CE marked in line with the EU medical device directives in order to ensure they are regulated and acceptably safe to use and also perform in the way the manufacturer/ developer intends them to. Health related apps and software that are not medical devices can be extremely useful but fall outside the scope of the MHRA.

When apps are not medical devices

Those apps that are not medical devices may be considered to be mHealth products. Work is ongoing at European level to determine a suitable legal framework.

The Concept of Mobile Health (mHealth)

Mobile health (hereafter "mHealth") covers *"medical and public health practice supported by*
mobile devices, such as mobile phones, patient monitoring devices, personal digital assistants
(PDAs), and other wireless devices"

It also includes applications (hereafter "apps") such as lifestyle and wellbeing apps2 that may connect to medical devices or sensors (e.g. bracelets or watches) as well as personal guidance systems, health information and medication reminders provided by sms and telemedicine provided wirelessly. mHealth is an emerging and rapidly developing field which has the potential to play a part in the transformation of healthcare and increase its quality and efficiency.

mHealth solutions cover various technological solutions, that among others measure vital signs such as heart rate, blood glucose level, blood pressure, body temperature and brain activities. Prominent examples of apps are communication, information and motivation tools, such as medication reminders or tools offering fitness and dietary recommendations. The expanding spread of smartphones as well as 3G and 4G networks has boosted the use of mobile apps offering healthcare services. The availability of satellite navigation technologies in mobile devices provides the possibility to improve the safety and autonomy of patients. Through sensors and mobile apps, mHealth allows the collection of considerable medical, physiological, lifestyle, daily activity and environmental data. This could serve as a basis for evidence-driven care practice and research activities, while facilitating patients' access to their health information anywhere and at any time.

mHealth could also support the delivery of high-quality healthcare, and enable more accurate diagnosis and treatment. It can support healthcare professionals in treating patients more efficiently as mobile apps can encourage adherence to a healthy lifestyle, resulting in more personalised medication and treatment.

It can contribute to the empowerment of patients as they could manage their health more actively, living more independent lives in their own home environment thanks to self- assessment or remote monitoring solutions and monitoring of environmental factors such as changes in air quality that might influence medical conditions. In this respect, mHealth is not intended to replace healthcare professionals who remain central to providing healthcare but rather is considered to be a supportive tool for the management and provision of healthcare.

21 CFR Part 11 -Electronic Records & Signatures

Part 11 of the FDA CFR is relevant to "records in electronic form that are created, modified, maintained, archived, retrieved, or transmitted under any records requirements set forth in Agency regulations." This first section of the book provides a background information and explanations of each section and requirement of the regulation. The second half of this eBook provides a clear and transferrable verification process for each requirement of 21 CFR Part 11, with suggested verification methods included.

As of 2007, several sections of the regulation have been identified as excessive and the FDA announced in guidance that it will exercise enforcement discretion on some parts of 21 CFR part 11. This has been welcomed by some manufactures but it has also causes a degree of confusion.

The requirements relating to access controls are the most fundamental requirements and are routinely enforced. The "predicate rules" that required organizations to keep records the first place are still in effect. If electronic records are illegible, inaccessible, or corrupted, manufacturers are still subject to those requirements.

If a regulated firm keeps "hard copies" of all required records, those paper documents can be considered the authoritative document for regulatory purposes. This then means that the computer system is not in scope for electronic records requirements, although subject to predicate rules which still require validation. If the "hard copy" is to be identified as the authoritative document, the "hard copy" must be a complete and accurate copy of the electronic source. The manufacturer must use the hard copy (rather than electronic versions stored in the system) of the records for regulated activities.

Definition of Records

The FDA has deemed the following records or signatures in electronic format subject to 21 CFR part 11:

"Records that are required to be maintained under predicate rule requirements and that are maintained in electronic format in place of paper format. On the other hand, records (and any associated signatures) that are not required to be retained under predicate rules, but that are nonetheless maintained in electronic format, are not part 11 records.

Records that are required to be maintained under predicate rules, that are maintained in electronic format in addition to paper format, and that are relied on to perform regulated activities. Records submitted to FDA, under predicate rules (even if such records are not specifically identified in Agency regulations) in electronic format (assuming the records have been identified in docket number 92S-0251 as the types of submissions the Agency accepts in electronic format). However, a record that is not itself submitted, but is used Contains Nonbinding Recommendations in generating a submission, is not a part 11 record unless it is otherwise required to be 205 maintained under a predicate rule and it is maintained in electronic format.

Electronic signatures that are intended to be the equivalent of handwritten signatures, initials, and other general signings required by predicate rules. Part 11 signatures include electronic signatures that are used, for example, to document the fact that certain events or actions occurred in accordance with the predicate rule (e.g. approved, reviewed, and verified)." The above definitions are taken from FDA guidance document entitled "FDA Guidance for Industry: 21 CFR Part 11 - Electronic Records and Electronic Signatures: Scope and Application, August 2003." This document also provides recommendations on documenting key decisions that may be taken in relation to 21 CFR Part 11 applicability and compliance.

Requirements and Specifications

The need for compliance to 21 CFR depends on type of technology and level of automation and computerisation involved in the manufacturing process or other actives that are GxP impacting. Does the system store electronic records? Does the system require a login? Is there an audit trial? If a complex system is to be procured, the requirements need to be communicated to the manufacturer as part of a User requirement specification and/or software requirement specification.

General Guidance on Requirement Specifications

While the Quality System regulation states that design input requirements must be documented, and that specified requirements must be verified, the regulation does not further clarify the distinction between the terms "requirement" and "specification." A requirement can be any need or expectation for a system or for its software. Requirements reflect the stated or implied needs of the customer, and may be market-based, contractual, or statutory, as well as an organization's internal requirements.

There can be many different kinds of requirements (e.g., design, functional, implementation, interface, performance, or physical requirements). Software requirements are typically derived from the system requirements for those aspects of system functionality that have been allocated to software. Software requirements are typically stated in functional terms and are defined, refined, and updated as a development project progresses. Success in accurately and completely documenting software requirements is a crucial factor in successful validation of the resulting software. Page 6 Guidance for Industry and FDA Staff General Principles of Software Validation A specification is defined as "a document that states requirements." (21 CFR 820.3(y)) It may refer to or include drawings, patterns, or other relevant documents and usually indicates the means and the criteria whereby conformity with the requirement can be checked. There are many different kinds of written specifications, e.g., system requirements specification, software requirements specification, software design specification, software test specification, software integration specification, etc. All of these documents establish "specified requirements" and are design outputs for which various forms of verification are necessary.

Validation of Computerised Systems

The requirement for computerised systems to be compliant to 21 CFR part 11, needs to be identified early on the project to ensure that the vendor or supplier of the systems or equipment can develop, build a system that meets the requirements of 21 CFR part 11. Computer system validation can be divided into 3 distinct phases which include: (1) Plan, (2) Design & Development, (3) verification and (4) Retirement. The requirement for computerised systems to be compliant to 21 CFR part 11, needs to be identified early on the project to ensure that the vendor or supplier of the systems or equipment can develop, build a system that meets the requirements of 21 CFR part 11.

Plan: This phase involves the planning of the validation effort required to implement the system and identification of key milestones and requirements. It requires supplier assessments, assessments of the regulatory and system risks, supplier, development of a validation approach and the identification of deliverables that will be generated, that will support the implementation and operation of the system.

Design & Development: This phase consists of the design, development and configuration of the hardware and software required to meet the system requirements. In case of custom software, design and developmental testing is important to ensure proper functionality prior to verification testing.

Verification: This phase confirms that requirements and specifications have been met. Testing is required to ensure the system operates as intended. Upon successful testing and verification, the system can be released for use. Once verification activities have begun any changes to the system must managed through change control. In case of successful completion of the verification activities (i.e. any deviation has been evaluated and addressed), the system is released for effective use. Operation This phase supports the need to maintain compliance and fitness for intended use after the system is accepted and released for use.

Retirement: This phase consists of the planning, executing and summarizing of the events required for system shutdown. It includes the appropriate handling of the supporting documents and the data contained within the system. While described here as a separate phase, a system's retirement can be handled as part of a new system implementation or as a separate project.

Best practice when it comes to Computer System validation is to adopt a life cycle approach for computer systems which requires the completed of activities in a systematic way from system conception to retirement. Life cycle activities could be scaled according to system impact on product quality, patient safety and data integrity, system complexity and novelty, supplier assessment and business risk.

Definitions

Computer System: A computer / automated system consisting of the hardware, software, and network components, together with the controlled functions (personnel, procedures, and equipment) and associated documentation.

Computer System Validation: A process that confirms by examination and provision of objective evidence that the computer system conforms to user needs and intended uses. Computer System validation is a process for achieving and maintaining compliance with GxP regulations and fitness for intended use by adoption of life cycle activities, deliverables, and controls.

GxP Regulated Computer Systems: Computer systems determined to have a potential impact on Product Quality, Patient Safety and Data Integrity; these systems are required to comply with the relevant GxP regulations.

Data Integrity: is the degree to which data is reliable and without error. Data must be accurate, attributable, contemporaneous, original, legible and available. A breach of data integrity occurs when any person manipulates or distorts data and submits the results of that data as valid.

Predicate rules: a predicate rule is any FDA regulation that requires companies to maintain certain records and submit information to the agency as part of compliance.

To gain a better understanding of the validation of computerized systems, consult the following publication- "FDA's guidance for industry and FDA staff General Principles of Software Validation." Industry guidance such as the GAMP 5 guide issued by ISPE is also a useful reference.

Electronic Records

When it comes to the regulated industries such as the medical device industry, every process and procedure must be documented. Documentation ensures that everyone is working in the same manner with the same procedures. However, documentation is more than just writing down procedures and processes. It is also concerned with how documents are controlled, how they are updated and how they are stored.

Electronic Document management systems

Electronic document management systems aka EDMS are now the norm and gold standard for most medium to large organisations. Many companies that provide medical device manufacturers with an EDMS can be customised to match the business processes particular to an organisation. With configurable or customisable software, validation and proper verification is important to ensure the system operates as intended. There are also regulatory requirements that stipulate the expectations and requirements of such system. For example, the application of electronic signatures and the presence of audit trials. FDA 21 CFR Part 11 details the requirements with regards to electronic records and electronic signatures. For medicinal products in Europe, GMP V4 Annex 11 specifies similar requirements.

Record Retention

Regard to the part 11 requirements for the protection of records to enable their accurate and ready retrieval throughout the records retention period (11.10 (c)) Persons must also comply with all applicable predicate rule requirements for record retention and availability such as (211.180(c) general requirements. The decision to follow 21 CFR part 11 should be justified and documented as part of a risk assessment and based on the value of the records over time.

FDA does not object to archiving of required records in electronic format to non-electronic media such as paper, or to a standard electronic file format (examples of such formats include, but are not limited to, PDF, XML, or SGML). Persons must still comply with all predicate rule requirements, and the records themselves and any copies of the required records should preserve their content and meaning. As long as predicate rule requirements are fully satisfied and the content and meaning of the records are preserved and archived, you can delete the electronic version of the records. In addition, paper and electronic record and signature components can co-exist as long as predicate rule requirements are met and the content and meaning of those records are preserved.

Electronic Signatures

Electronic signatures are computer-generated character strings that count as the legal equivalent of a handwritten signature. The regulations for the use of electronic signatures are set out in 21 CFR Part 11 of the FDA. Each electronic signature must be assigned uniquely to one person and must not be used by any other person. It must be possible to confirm to the authorities that an electronic signature represents the legal equivalent of a handwritten signature. Electronic signatures can be biometrically based or the system can be set up without biometric features.

Conventional Electronic Signatures

If electronic signatures are used that are not based on biometrics, they must be created so that persons executing signatures must identify themselves using at least two identifying components. This also applies in all cases in which a chip card replaces one of the two identification components. These identifying components, can, for example consist of a user identifier and a password. The identification components must be assigned uniquely and must only be used by the actual owner of the signature.

When owners of signatures want to use their electronic signatures, they must identify themselves by means of at least two identification components. The exception to this rule is when the owner executes several electronic signatures during one uninterrupted session. In this case, persons executing signatures need to identify themselves with both identification components only when applying the first signature. For the second and subsequent signatures, one unique identification component (password) is then adequate identification.

Audit Trail

Title 21 CFR details predicate rule requirements relating to documentation of, for example, date time, or the sequencing of events, as well as any requirements for ensuring that changes to records do obscure previous entries.
Making the decision on whether to apply audit trails, or other appropriate measures, or on the need to comply with predicate rule requirements should involve a justified and documented risk assessment. Any Risk assessment should determine the potential effect on product quality and safety and the integrity of the record.

Change Management

Validation programs are subject to change control. Each company or organisation should have a procedure detailing the change management process. Below is a suggested overview of a typical change control process. Any system, facility, document or process that has the potential to impact product quality and validated state is generally subject to following a change control process. Another term used in industry is Enterprise Change Control or Engineering Change Control. Essentially these terms are the same. The intent is to control and manage change consistently.

A change control can take the form of a document which drives the agenda and the specific requirement. Change control is also created with enterprise software such as Kintana, Documentum and SAP. While each company will have varying processes, some basics are common. These include the 3 stages of change control; pre-implementation, implementation and post implementation (if required). Below, 2 case studies are detailed where there is a change in manufacturing which requires a formal change control process to be applied.

Design for Manufacture

Design for Manufacturability (DFM, (design for manufacturability) essentially is the practice of designing products in a manner that fosters efficient and reliable manufacturing methods. The core benefit and motivation is to reduce costs. However, this is achieved in numerous ways.

1) Reduce the number of parts

2) Use of modular design

3) Reduce handling

4) Remove product features not beneficial to the patient or required by regulations

5) Select materials that can be processed easily on the manufacturing equipment

6) Simplify packaging systems while ensuring compliance to safety and quality

7) "Built-in" Quality from the beginning

8) Eliminate compliance errors

A DFM mindset can work to ensure that potential problems are fixed and resolved in the design phase which is the least expensive phase in the product lifecycle to provide solutions.

Other factors may affect the manufacturability such as the type of raw material, the form of the raw material, dimensional tolerances, and secondary processing such as finishing.

Depending on various types of manufacturing processes there are set guidelines for DFM practices. These DFM guidelines help to precisely define various tolerances, rules and common manufacturing checks related to DFM.

While DFM is applicable to the design process, a similar concept called DFSS (Design for Six Sigma) is also practiced in many organizations.

1. Reduce the number of parts

- Reducing the number of parts in a product inherantly reduces complexity and the potential need for changes to dimensions and features. If changes are required, a quicker design tunraround can be achieved.

2. Use a Modular Design

- "Modules" in product design simplifies manufacturing such as inspection, assembly, testing and redesign etc. A modular product essentially divides a product into several parts that can be working on individually. In some respects, a car or vehicle is modular. Different components are manufactured or machined independently and then come together to perform as one. Medical implants such as orthopaedic hips and knee replacements commonly use modular designs.

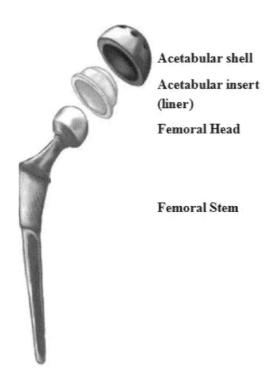

Acetabular shell

Acetabular insert (liner)

Femoral Head

Femoral Stem

Figure: A total hip replacement (THR) consisting of a modular design. The acetabular insert can be manufactured and inspected independently of the acetabular shell. The insert may have more critical measurements and features than the shell thus allowing the shells to be discarded if they fail to meet quality requirements.

3. Reduce handling

- Handling can include human/operator handling or automated handling by a machine or robot. Handling is required in manufacturing in oder to position or present a part in a certain orientation or position. A simple way of reducing handling is to design symmetrical parts. Jigs are fixture can also be utilised to reduce handling times by providing a fixture to allow east orientation

4. Remove obsolete features

- Some product may have features built in to the design of a product aiming to enhance the functionality or appearrance of a product. However, a good rule of thumb is to question if the feature is prescribed by quality, safety or regulation authorities, or if it is a user or patient requirement.

5. Materials Selection

- Using Standard components is less expensive than bespoke/custom alternatives. This can simplify manufacturing in terms of availability of parts and associated lead times.

6. Simplify packaging

- Standardisation of pack size ad equipment is an effective way to cut out complexity and the problems associated with custom packaging designs.

7. "Built-in" Quality

- Quality by Design

8. Eliminate compliance errors

- Compliance errors occur during insertion and assembly operations as a result of varying dimensions in components. Another cause of these errors is the accuracy of the positioning device (manual, automated, template etc)

Jigs and Fixtures

The terms "jig" and "fixture" are commonly used in the manufacturing industry, particularly in CNC machining and fabrication. Many machining processes require jigs and fixtures in order to achieve consistent and accurate outcomes which in turn contributes towards a DFM philosophy.

Jig: a jig is used to guide the item or component that has to be machined (e.g. CNC turning) or reversely if the component is stationary while a fixture holds in place of "fixes" the component to be machined or processed.

Fixture: a fixture is used to hold the component or part during the machining process. Its purpose is not to guide the part towards the machining tool. Fixtures are secured with the table surface of the mills in most of the cases. Fixtures reduce the need for other tools and facilitate more accurate machining and processes.

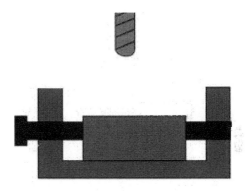

Fixture (Above) Secures workpiece

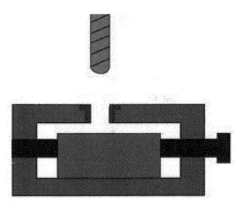

Jig (Above) Secures workpiece and allows operator to accurately position hole.

Design for ease of fabrication ensures that makes are avoided or eliminated. This starts with the correct materials and the use of correct assembly options. Select the optimum combination between the material and fabrication process to minimize the overall manufacturing cost. In general, final operations such as painting, polishing, finish machining add to overall cost. Excessive tolerance, surface-finish requirement, and so on are commonly found problems that result in higher than necessary production cost.

Made in the USA
Middletown, DE
27 September 2022

11352492R00061